SHOES THAT DON'T HURT

DANIEL A. FRIED

Disclaimer: The contents of this book, such as text, graphics, images, and other material are for informational purposes only and are not intended to offer medical advice.

Note for Librarians: A cataloguing record for this book is available from Library and Archives Canada at www.collectionscanada.ca/amicus/index-e.html
ISBN 1-4120-9706-1

Printed in Victoria, BC, Canada. Printed on paper with minimum 30% recycled fibre.
Trafford's print shop runs on "green energy" from solar, wind and other environmentally-friendly power sources.

TRAFFORD
P U B L I S H I N G
Offices in Canada, USA, Ireland and UK

Book sales for North America and international:
Trafford Publishing, 6E–2333 Government St.,
Victoria, BC V8T 4P4 CANADA
phone 250 383 6864 (toll-free 1 888 232 4444)
fax 250 383 6804; email to orders@trafford.com
Book sales in Europe:
Trafford Publishing (UK) Limited, 9 Park End Street, 2nd Floor
Oxford, UK OX1 1HH UNITED KINGDOM
phone 44 (0)1865 722 113 (local rate 0845 230 9601)
facsimile 44 (0)1865 722 868; info.uk@trafford.com
Order online at:
trafford.com/06-1462

10 9 8 7 6 5 4 3 2

To my mother, Blanche Fried

and the memory of my father, Emanuel Fried

Contents

Illustrations

I'm having a good time tonight. If nobody objects, I'm going to take off my shoes.

Joan Baez

Introduction

This book is written for a particular reader: one who knows that wearing shoes can be a problem and is willing to read through some technical (but not difficult) material to gain an understanding of the problem and its solutions. Many books have been written treating shoes as an item of fashion, that is to say, an object to be looked at. I offer another perspective: shoes as a medium through which forces are exchanged between a human being and the Earth. The forces are the ones described in Newton's laws of motion. The exchange of forces is what enables us to move across the Earth's surface through the activities of walking and running. One of the premises of this book is that if we understand the Newtonian forces that pass through shoes when we walk, we will have a foundation for understanding how to make shoes that do not create problems for the human body.

This book is a product of my experience in wearing, thinking about, and making shoes. I had a long struggle with the shoe industry in trying to find shoes that were comfortable to walk in. I could not wear ready-made shoes because my feet were just a little too wide in the front. I could jam my toes into the shoes, but it was painful. I therefore turned to custom shoemakers. There was a small community of them in New York a few decades ago, most, if not all European-trained. Their craftsmanship was impeccable, but I was still not comfortable walking in their shoes. Since I had no problems when walking barefoot, I had to conclude that there was a flaw in the basic design or structure of conventional shoes. It's obvious when you look at a man's shoe, and even more so at a woman's shoe, that any heeled shoe sets the foot in an unnatural position, and thereby restricts the movement of the foot to some extent. This might not be a problem for some people, but it was evidently a problem for me.

It is a widely held dogma that shoes have to "support" the feet. I doubted this on simple anthropological grounds: human beings walked the earth for ages without anything resembling contemporary shoes. I could see no reason why a minimalist approach would not work. That meant a shoe that did little more than cover the foot, letting the foot do all the work. For a few years, I put this philosophy into practice by wearing Indian-style moccasins. These were hand made from leather, using a simple pattern of the type that can be found in leather craft books. I coated the bottoms with a sneaker repair substance so that they could be worn outdoors without being abraded through by sidewalk concrete. I walked all around New York City in these moccasins. In many ways, they were more comfortable than the most expensive custom men's shoes. There were problems, however. A thin piece of leather provides no shock absorption. This is bad news when you're walking on sidewalks. There was also no thermal insulation. In winter, the cold from the freezing sidewalks went right through the sole. It was the same in the summer with the heat from the sunbaked streets. For these reasons, I eventually gave up on the moccasins and bought a new pair of custom shoes.

The moccasin experience convinced me that what I really needed was shoes with the heellessness and flexibility of moccasins, but with adequate shock absorption and thermal insulation, and with the general appearance of regular shoes. I thought I could make something like it if only I could learn how to make shoes. Considerations of artistry aside, there is little more to making moccasins than tracing a pattern and cutting and stitching the leather. Shoemaking, however, is a technology. It has to be learned, but it was not feasible to apprentice myself to a custom shoemaker. What came to my rescue was the Internet. I had reached the end of the road with my post-moccasin shoes and was searching the web for custom shoes. I came upon the web site of ShoeSchool.com. The school is located on the Olympic Peninsula in Washington State and is run by Alan Zerobnick, a veteran shoemaker and shoe industry consultant. He offered an inspired four-day workshop (later expanded to five days) in the essentials of hand made shoes. With very little hesitation, I signed up.

The course gave me the impetus to resume making footwear. I now had enough knowledge to make a basic pair of shoes. With help from a sculpture supplies store (another New York City amenity), I was able to make custom lasts based on a cast of my feet. I began putting together shoes according to my long-held ideas. I used some materials similar to what I had seen in running shoes. After some time and experimentation, I was able to achieve a satisfactory result: shoes that had no heel, were flexible, had enough shock absorption and thermal insulation for walking on city streets, and looked like shoes. They did not have the finish and workmanship of professionally made shoes, but they were much more comfortable.

All the while that I was making shoes, I kept thinking about the underlying science. With a minimalist design philosophy, it is important to be clear about what the real minimum is that shoes have to do. That clarity can only come from an understanding of the relevant scientific principles. And the science is interesting in its own right.

Putting together the trial-and-error experience and the science, I have sought to identify the basic features that make a minimalist shoe work in an urban environment. Collectively, these features constitute what I call the "isodynamic shoe design" – "iso" from the Greek word for equal and "dynamic" meaning pertaining to force producing motion. The name expresses the idea that the feet should be applying the same set of forces when walking

3

in shoes as they do when walking barefoot. The isodynamic design is not a radical idea. In some measure, it is simply an optimization of parameters that have existed in footwear since the earliest prehistoric moccasins.

There are two principal themes in this book. One is the isodynamic design. It is briefly introduced in Part 2. Part 3 gives a detailed analysis of its structures and materials, presented in the context of how the design achieves its desired goals. The second theme is broader. It is an inquiry into the question of what shoes should do. This theme occupies Part 2 and culminates in a list of performance specifications in Chapter 8. The two themes overlap in practice but are analytically distinct. The second theme challenges the reader to think about the basic functions of shoes. The isodynamic design is offered as one possible way of fulfilling those functions. The science in Part 1 lays the groundwork for the ideas presented in Parts 2 and 3. Part 4 examines various parts of the shoe. The design features discussed in Part 4 are not essential elements of the isodynamic design, but they play a great role in whether or not a shoe hurts. Although I have arranged the four parts of the book in logical order, working from the general to the particular, the reader may find it more interesting to start with Parts 3 and 4. This is where the actual design features will be found.

I may be criticized for writing this book without being a professional in the fields of physiology or footwear design. My answer to this is that I do not ask the reader to accept anything in this book on the strength of my authority. Rather, I ask the reader to think critically about the ideas contained in the book. Many of them derive from elementary science and common experience. The ultimate proof of the validity of the isodynamic design is how well the shoes perform in practice. My own experience with them has been entirely satisfactory. Time constraints have prevented me from making shoes for more than a small number of other people. I would be happier if I could cite a wider experience with the shoes, but this book might never see the light of day if I had to wait until that was possible. I believe the issues raised in this book are substantial enough to warrant its publication at this time.

I invite the reader now to sit back, take off your shoes, and get ready to think about walking the Earth under the reign of Newton's Laws of Motion.

New York City
May 2006

Part 1

Science of Human Movement

To fully appreciate how shoes function, we need to know something about human locomotion, and for this we consult two sciences: physics and biology. In physics, we look at some basic laws of Newtonian mechanics. In biology, we look to biomechanics for an understanding of the structure of human motion, and neuroscience for an understanding of the kinesthetic sense. The brief review of physical and biological principles in this Part will give us a good scientific perspective on the enterprise of shoe design.

Chapter 1

Newton's Laws of Motion

Walking is a process by which one body (a human being) changes its spatial position relative to another body (the earth). This process is subject to Newton's laws of motion.

Newton's First Law states that "[e]very body perseveres in its state of rest or of uniform motion in a right line unless it is compelled to change that state by forces imposed thereon." The Second Law states that the acceleration (change of motion) of a body is directly proportional to the force applied to it and inversely proportional to its mass. This is expressed in the formula F=ma. The Third Law states that "[t]o every action there is always opposed an equal and opposite reaction." (Encyclopedia of Physics, p. 811.)

These three laws have profound implications for the act of walking. The First Law tells us that you can't generate motion by yourself. That is, you can't change the position of your center of mass relative to the earth's. A force outside your body has to push or pull you.

The Third Law explains where the outside force comes from. It comes from the reaction of the earth to being pushed by your feet. As Aristotle observed, "the animal that moves makes its change of position by pressing against that which is beneath it" (Aristotle, p. 489). Your muscles generate forces that push against the earth. The earth pushes back. The forces exerted by the earth (or ground) against the body are referred to as "ground reaction forces" (Thompson). Figure 1 is a simplified vector diagram of the forces exerted by the body and the ground while walking forward. The vector lines represent *net* forces exerted over one or more complete gait cycles (as defined in Chapter 2). AF(h) and AF(v) are the net horizontal and vertical forces *applied* by the body against the ground. GRF(h) and GRF(v) are the net ground reaction forces in the same planes. (Side-to-side forces are ignored.) The ratio of the v vectors to the h vectors is arbitrarily selected. The diagram illustrates that the net horizontal force exerted by the feet is retrograde, *i.e.*, opposite, to the person's direction of motion. This is necessary in order to generate a net forward ground reaction force.

The Second Law explains how fast you move. The acceleration of your body at each step is a function of the strength of the ground reaction force and your mass (weight).

Shoe design must take into account the fact that the bottom of the shoe is a medium through which forces are exchanged between the foot and the ground. Of particular interest in the isodynamic design is the fact that the foot must push backward against the ground in order to generate forward motion. We will see the consequences of this later on.

From a Newtonian viewpoint, then, the act of walking is a process of generating the ground reaction forces necessary to move a person's body. The way that the body generates those forces is explained by the biomechanics of walking.

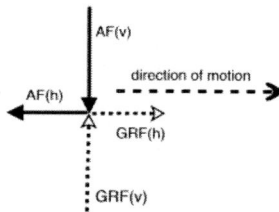

Figure 1. Vector diagram of net applied forces and ground reaction forces. The net ground reaction force is upward and in the direction of motion.

Chapter 2

Biomechanics of Walking

The structure of the human body determines the precise manner in which we push against the earth to generate ground reaction forces. Biomechanics is a complex subject: this chapter will consider a few of the biomechanical phenomena that are of concern in shoe design.

The Gait Cycle

The act of walking can be seen as a recurring cycle of discrete events. The gait cycle has been defined as "the time interval between two successive occurrences of one of the repetitive events of walking" (Whittle, p. 58). We follow the right foot through a single cycle, divided into seven "periods." You can walk this off yourself as you read. A person is already in motion when the first period begins.

The *first* period is "initial contact," also known as "heel strike." It begins when the heel of the right foot strikes the ground after the right leg has swung forward past the left leg. The *second* period is "opposite toe off." It begins when the left toe leaves the ground. The *third* period is "heel rise." It begins when the right heel begins to lift off the ground as weight shifts to the forefoot. At this point, the left foot is moving forward past the right leg. The *fourth* period is "opposite initial contact." It is the initial contact of the left foot. The *fifth* period is "toe off." It begins when the right toes leave the ground. The *sixth* period is "feet adjacent." It begins when the right foot is next to the left foot, the right leg being in the act of swinging forward. The *seventh* period is "tibia vertical." It begins when the tibia of the right leg is in a vertical position as the right leg swings forward. The seventh period is followed by initial contact as the cycle repeats (pp. 71-89).

9

The four periods from initial contact to opposite initial contact constitute the "stance phase" of the gait cycle. The remaining periods constitute the "swing phase" (p. 59).

The power stroke of the gait cycle occurs during the heel rise and opposite initial contact periods, when the forefoot is pressing hardest down and back against the ground. This action may be referred to as "push-off" (p. 131). Figures 4 and 12 illustrate the position of the foot during push-off. There are strong retrograde forces exerted by the foot during push-off. These forces outweigh the forward (braking) forces exerted against the ground by the heel in the initial contact period, resulting in a net retrograde force over the complete cycle. The net retrograde force is what generates a net forward ground reaction force over the cycle.

Transverse and Oblique Axes

The foot has two axes of flexion in the forefoot--the transverse and oblique metatarsophalangeal axes (Zatsiorsky, pp. 292-295). To see this, look down at your bare foot. Imagine a straight line from base of the large toe to the tip of the small toe. This is the transverse axis. Now imagine a straight line from the base of the small toe to the base of the second toe, continuing through the large toe. This is the oblique axis. The foot is capable of flexing about either of these axes during push-off. When walking at a moderate or fast speed, or running, it is normal to flex and push off about the transverse axis. When starting to walk, stopping, squatting or carrying a heavy load, it is advantageous to flex about the oblique axis. The difference between the two axes has been likened to the difference between high and low gears (p. 293).

Toes and Arch

Hyperextension, *i.e.*, bending back, of the toes, facilitates raising of the arch (Zatsiorsky, pp. 294-295). You can demonstrate this by placing one foot flat on the floor and raising the heel while keeping the toes flat on the floor. During barefoot walking, such hyperextension occurs prior to toe off. There is also a hyperextension that occurs during the swing phase. These can be clearly seen in Eadweard Muybridge's photographic studies (Muybridge, Plate 3). Shoes that prevent the toes from fully hyperextending can cause a permanent deformity known as *hallux valgus* (Zatsiorsky, p. 294).

Rotation of the Foot

If you imagine a primitive robot, it probably has one-piece feet that rotate on hinges at the bottom of the legs. The human foot, by contrast, moves in a much more complex way during the gait cycle. Among other things, the foot rotates about its long axis. You can get an idea of this by putting your foot on the ground and rolling it from side to side. The rotation is limited: you can't spin your foot around the way you can spin your hand. The rotations of the foot are coupled to its other movements: raising your heel while the toes remain on the ground causes the hindfoot (the part under the ankle) to rotate outward. Finally, the foot does not rotate as a unit while walking: the forefoot, midfoot and hindfoot rotate differently at different points of the gait cycle. (Whittle, pp. 70-89; Zatsiorsky, pp. 299-300.)

Range of Motion

The "range of motion" of a joint is the difference between the two extremes of joint movement (Zatsiorsky, p. 227). For example, the range of motion of the elbow is the difference between the completely flexed (bent) position and the completely extended (straight) position. Moving the elbow "through" its range of motion means flexing and extending the arm so that the elbow goes from one end of its range to the other.

Flexibility in the human body can be defined as "a measure of the range of motion available at a joint or group of joints" (Surburg, p. 102). Flexibility can be lost over time if a joint is not regularly exercised through its full range of motion (Watkins, p. 137). One of our concerns in shoe design is whether the shoe restricts the range of motion of any joint.

Chapter 3

Kinesthetic Sense

K inesthetic sense is the conscious and unconscious awareness of body position and body movement (Watkins, p. 236). One component of kinesthetic sense is proprioception, which is defined in Stedman's Medical Dictionary as "the unconscious perception of movement and spatial orientation arising from stimuli within the body itself." Proprioception aids in coordination and balance (Watkins, pp. 236-240).

Reduction of sensory awareness through the feet, resulting from disease or aging, leads to reduction of standing and walking balance (Collins). This is the principle behind the experimental use of vibrating soles to help improve the balance of people with neurological damage from stroke or diabetes (Priplata). The vibration increases sensory awareness.

Proprioception is a stream of information. When a person is walking or standing, the position and motion of the feet, ankles and legs enter that stream. Sensory input also comes from the pattern of pressure exerted by the ground against the bottom of the feet in the course of a stride. Shoes can reduce kinesthetic sense by restricting the motion of the feet, ankles and legs, and by reducing the complexity of the foot's pattern of contact with the ground.

Part 2

Specifications

Shoes are made to be worn and walked in. To effectively judge how well a pair of shoes fulfills these purposes, we need a list of the things that shoes should do: performance specifications. In this Part, we develop a set of specifications for shoes. We begin by considering the anatomy and function of shoes. We then examine the concept of an ideal walking surface. Those topics will lay the foundation for our specifications. In the chapter on shoe anatomy, we get a first look at the isodynamic shoe design.

Chapter 4

Assumptions

A shoe design can be properly evaluated only in the context of its typical use. In this book, I make certain assumptions about the environment in which the shoes are worn and the people who wear them.

My paradigm for the use of shoes is walking on city sidewalks. It is likely that many of the design principles discussed in this book apply to running and other athletic activities, and in a broad range of environments, but that is beyond the scope of this work.

I assume that the wearer does not have a condition requiring the use of corrective shoes or orthotics, or any other device that restricts motion of the feet.

The basic design principles apply to men and women, children and adults. Some details of construction will vary according to weight and other individual characteristics.

Chapter 5

Anatomy of Shoes

This chapter introduces some basic shoe anatomy and terminology. I will start with what I refer to as the "modern shoe," and follow with the isodynamic shoe. Unless otherwise noted, I use standard terminology of the footwear industry as defined in The Complete Footwear Dictionary. Note that when I speak of "design," I mean the engineering sense of the word. By "shoe design," I mean the general construction or architecture of the shoe. This is not the same thing as "shoe style." A shoe style is an amalgam of elements relating to the shoe's appearance as well as its architecture.

Figure 2 is an illustration of a men's dress shoe, identifying some of its principal parts. The style of this shoe is usually called an *oxford* although, more precisely, it is a *blucher*. (Readers interested in the distinction can consult The Complete Footwear Dictionary or Handmade Shoes for Men.) The design (in the engineering sense) is what I call the "modern shoe." It is modern in that it is only a few centuries old (Peacock, p. 7), in a history of footwear spanning at least twelve millennia. It is modern, also, in that it's the basic design of most contemporary styles of men's shoes and many styles of women's shoes. I use the

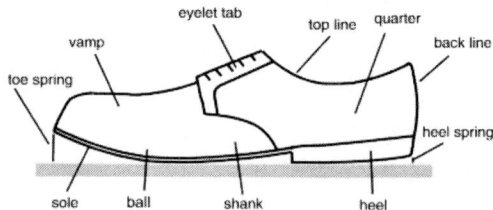

Figure 2. Modern shoe, exemplified by men's oxford.

men's oxford style for illustration, partly because it's what I am most familiar with, and partly because of its universality across men's and women's shoes.

The modern shoe is, with some modifications, the basic design of most athletic shoes on the market today. By "athletic shoes," I mean running shoes and other sports shoes with a relatively rigid bottom and back. This does not include sneakers, because sneakers are flexible.

The defining features of the modern shoe are:

1.   a raised heel;

2.   an upward slope from around the middle of the heel to its back edge, known as the "heel spring";

3.   a downward angle from the front (breast) of the heel to the ball tread area. This part of the shoe is known as the "shank." It is often reinforced with a piece of material such as steel or fiberglass.

4.   an upward slope from the ball to the tip, known as the "toe spring";

5.   a construction (geometry and materials) in the front sufficiently rigid so that the curve of the sole does not straighten when walking;

6.   a construction in the front sufficiently flexible so that the front of the shoe can bend inward;

7.   rigid materials and reinforcements in the back part of the upper, known as "heel counters," so that back part maintains its shape throughout the gait cycle.

The modern shoe was created at a time when leather and wood were the only practical materials for virtually all parts of the shoe. From an engineering perspective, it may be seen as a solution to the problem of walking on hard surfaces such as streets and floors. It was a good solution given the constraints of then available materials.

Figure 3 is a basic isodynamic shoe. The style is an oxford (or blucher). Like the modern shoe, it can be made in other styles. In contrast to the modern shoe, the isodynamic shoe is flat along the bottom. There is no raised heel. Therefore, there is no toe spring, shank or heel spring. The materials and construction allow the sole and heel to bend with the flexion and extension of the walking foot. The backpart (the part of the quarter extending back from the front of the heel) is soft. With use, the bottom may curl up somewhat in the front and back, but the bottom retains its uniform thickness. The curl will be evident when the shoes are off; when the wearer stands in them, the shoes flatten out.

In the modern shoe, there is no single part called the "sole." Rather, there is an "insole," an "outsole," and possibly a "midsole." The insole is the piece of leather or other material to which the upper is attached. It extends over the entire length of the shoe. The upper surface of the insole comes into contact with the foot, unless the insole is covered by a lining or an insert. The outsole is the very bottom layer of the shoe, from the tip of the toe to the heel breast, that comes in contact with the ground. There may be a midsole between the outsole and the portion of the insole ahead of the heel breast. The midsole is normally thicker than the outsole and its main function is cushioning or shock absorption. The modern shoe depicted in Figure 2 has no midsole. This is typical of a men's dress shoe. The isodynamic shoe has a midsole. In Figure 3, the bottom line represents the outsole and the white space above it is the midsole.

In the isodynamic shoe, the sole is the anterior portion (approximately two-thirds) of the bottom. The heel looks like continuation of the sole, but its midsole is a separate piece of material from the sole's midsole. The outsole can be a single continuous sheet of material spanning the entire length of the shoe. The insole is the same as the insole of a modern shoe.

A few more notes on terminology:

When I refer to the "bottom" of the shoe, I mean the totality of the insole (plus linings), midsole, outsole and heel; in other words, the parts of the shoe that are underneath the foot when a person is standing or walking.

"Instep" refers to a part of the foot and the part of the shoe that covers it. Exactly which part of the foot varies among dictionaries, but commonly it means the top part of the foot in front of the ankle. Where the definitions vary is in how far forward the instep extends. I use "instep" to indicate the area of the foot approximately underneath the eyelet tabs and tongue in an oxford or blucher style, as shown in Figures 2 and 3. This is roughly equivalent to the midfoot, which is the part of the foot between the ball and the heel bones. In the isodynamic design, the instep is the one area of the shoe that must fit snugly.

The "back seam" is the seam that joins the two quarters of a shoe in the back in many shoe styles. It normally runs along the back line.

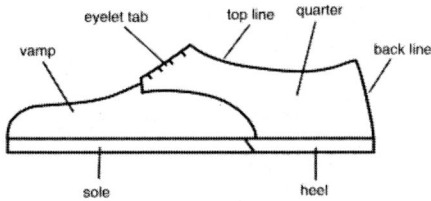

Figure 3. Isodynamic shoe.

Chapter 6

Functions of a Shoe

The first step in developing performance specifications for shoes is to ask what it is that shoes do. In other words, what are the functions of a shoe? As I see it, there are four basic functions.

The first function is to provide a walking surface. When you walk in shoes, the bottom of the shoe becomes your walking surface. You can think of it as two rows (left and right) of such surfaces, each surface separated from the previous one by the length of your stride. To state it differently, shoes are a portable road.

The second function is to protect the feet from the environment. This essentially means protection from heat, cold, water, and sharp objects underfoot.

The third function is to provide an environment for the feet within the spatial confines of the shoe.

The fourth function is esthetic. Shoes should look good on the person who wears them. The esthetic function includes everything under the heading of "fashion."

These are the general functions of shoes. There are special functions that shoes may have, specific to the wearer, the environment or a special use, but that is outside the scope of this book.

Chapter 7

Walking Surfaces

One of the basic functions of shoes is to provide a walking surface. In a sense, a pair of shoes *is* a walking surface; one that is carried on the feet and laid down with each step the person takes. We can gain a great deal of insight into shoe design by identifying what constitutes an ideal synthetic surface for barefoot walking. That surface is, more or less, what the bottom of a shoe should be. I say "synthetic" surface because the shoe bottom is a synthetic object. We can't expect to duplicate the complexity and variety of the earth's natural surfaces in a shoe.

By "walking surface," I do not mean the two-dimensional object of plane geometry. A walking surface in this context means a three-dimensional object that constitutes or covers the ground or floor. A carpet is an example of such a surface.

The discussion of walking surfaces in this chapter is organized around what I consider the four cardinal properties of a walking surface--hardness, resilience, friction and stability. I can't presently give precise optimal values for these properties. Indeed, there probably is no precise optimal value. The optimum varies from one person to another, and for each person it probably varies from one hour to the next. The picture is further complicated by the need for tradeoffs between the properties. The values I will propose for these properties are the ones that I have found to work in practice.

Hardness and Resilience

Hardness is a function of displacement. If you slam your fist down on a wooden table, there is virtually no displacement of wood. The surface does not perceptibly move. It's a hard surface. Your fist hurts when you hit that surface. If you do the same thing to a piece of foam rubber, the surface deforms and material is displaced. Your fist sinks down a certain distance into the material and it doesn't hurt. That's a soft surface.

In industry, the hardness of materials such as rubber and plastic is measured on various "Shore" hardness scales using an instrument called a durometer. The durometer measures the penetration of a specified indentor into the material under specified conditions of force and time. The Shore A scale is used for softer materials and the Shore D scale is used for harder ones.

Resilience is a correlative of hardness. Resilience is defined as "the property of a material that enables it to resume its original shape after being bent, stretched or compressed." Foam rubber is an example of a resilient surface. Dry sand is nonresilient. In a very hard material, resilience does not come into play, because the surface it is not significantly deformed. A resilient material stores part of the energy that goes into deforming its surface, and uses it to return the surface and underlying material to their original shape.

With these definitions in mind, let us look at how the properties of hardness and resilience work in walking surfaces. We will see that they have an important influence on ground reaction forces, kinesthetic sense, and shock.

Ground Reaction Forces

A walking surface must have a minimal degree of hardness to make walking possible. If you try to walk on water or quicksand, your feet will go all the way through the surface without stopping. The force exerted by your feet is used up in displacing the medium, and virtually none of it goes into pushing the Earth. No ground reaction forces are generated.

At the other extreme, think of walking on a concrete structure that is anchored to bedrock. That surface is hard and nonresilient. Virtually all the force exerted by the bottom of the feet goes into pushing (or spinning) the Earth. The ground reaction forces are almost equal to the applied forces.

Between the extremes are surfaces with varying degrees of hardness and resilience. Resilience is always less than 100%, so some energy will be lost to heat. For purposes of efficiently generating ground reaction forces, the harder and less resilient the surface, the better.

Kinesthetic Sense

Hardness has a positive effect on awareness. On a hard surface, you feel the ground as soon as your foot touches it, instantly giving you an awareness of the position of your foot. As the surface becomes softer, the sensory impression of the ground becomes less precise. In part this occurs because the surface is moving (deforming) under the foot, so the ground has, in effect, a less precise shape.

Awareness of the ground is a central part of the experience of barefoot walking. Loss or reduction of balance results when sensory awareness of the ground is diminished beyond a certain point by aging and disease (Collins). There is evidence that shoes with soft bottoms can diminish awareness to the point of causing harm. For example, it has been hypothesized that the reduction of sensory feedback is a cause of injuries in runners who wear running shoes as opposed to barefoot runners (Driscoll, p. 3). Also, it has been found that "shoes with thin, hard soles provide better stability for men than those with thick, soft midsoles" (Robbins, 1997) and that a low resilience interface improves stability in comparison with no resilience or high resilience (Robbins, 1998).

23

These factors argue in favor of a hard walking surface. However, the advantages of a hard surface must be balanced against the risk of shock.

Shock

The harder the surface, the sharper the impact. The bottom of the foot has some built-in shock absorption. The skin, subcutaneous fat, and muscles all contribute. The bursal sac in the heel gives added resilience on heel strike. Nevertheless, walking barefoot on a very hard surface, like concrete, can cause heel pain, and eventually, perhaps, injury. It is therefore prudent for a walking surface to provide some shock absorption.

I have found that a high-density EVA foam, with a Shore A hardness of approximately 45 and resilience of 25%, greatly reduces the shock of heel strike without unduly reducing the awareness of impact. This represents a good tradeoff between generation of ground reaction force, proprioception and shock absorption. To give the reader an idea of how hard this material is, if you press the edge of a fingernail into it, you will feel resistance, but the fingernail will make a mark in the foam; if you press the foam with the fleshy part of your fingertip, it will not make an impression.

Friction

Friction is the resistance of the motion of one object tangentially against another. Under Newton's laws of motion, the feet must exert a backward force against the ground in order for a person to walk forward. The heel exerts a forward force when it strikes the ground. Without friction, the forefoot would simply slide back against the ground during push-off and the heel would slide forward during heel strike. Little or no ground reaction force would be generated, and the person would struggle not to fall. Thus friction between the bottom of the foot and the walking surface is essential to walking.

Friction is measured in terms of the coefficient of friction (COF) between two designated objects or materials held together by a normal force. The greater the friction, the higher the COF.

If a person walks barefoot on any solid, dry surface, there is normally enough friction to avoid slipping. The feet, under the pressure of the person's weight, obtain a sufficient grip on the ground. Nevertheless, it may be easier for at some people to walk on surfaces with a high COF. They may find that on lower-COF surfaces, they have to bear down a little harder on the ball of the foot, and this may result in sprains in the feet, hips or elsewhere.

Stability

It is easier, in general, to walk on a surface that doesn't move around when the foot is bearing down on it. This would hardly be worth mentioning, except that a moving surface is what the modern shoe provides. The shoe pivots around the rear of the heel and the front of the sole. In the isodynamic shoe, in contrast, the heel and sole lie down under the weight-bearing part of the foot, without pivoting.

To summarize, the ideal walking surface is hard enough to maintain efficient energy transfer and kinesthetic sense, but resilient enough to provide shock absorption. It has a sufficiently high coefficient of friction to keep the feet from sliding. It is stable under the parts of the foot that bear weight during the gait cycle.

Chapter 8

Specifications

Having considered the functions of a shoe and the physics and biology of walking, we are ready to make a list of specifications for general purpose walking shoes. They are as follows:

1. Provide an optimal walking surface in terms of hardness, resilience, friction and stability.

2. Allow full range of motion of all parts of the body, particularly the foot and ankle.

3. Be neutral. In the chapter on biomechanics, we saw that there are two ways the foot can flex in the push-off phase of the gait cycle: it can flex about the transverse axis or the oblique axis. A neutral shoe does not favor one of these maneuvers over the other. It does not force, or even prompt, the foot in any particular direction. The foot does what it would do if it were bare. In general, a neutral shoe does not favor any one action of the foot over any alternative action. It does not impose its own pattern or rhythm on a person's gait.

4. Optimize kinesthetic awareness.

5. Protect the feet from the outside environment.

6. Provide a healthy internal environment for the feet.

7. Fit well.

8. Look good.

You may find it surprising that I have not included support or comfort in the list of specifications. I will explain my reasons in Part 3.

Chapter 9

Flat Shoes

The isodynamic shoe, as you see in Figure 3, is flat. The bottom is of an even thickness from front to back. But is a flat shoe better than a heeled shoe? Within the context of minimalist shoe design, the answer is clearly "yes." Putting a raised heel on a shoe changes the position of the foot and almost inevitably imposes some restriction on range of motion. These in turn may necessitate arch supports, toe spring and other apparatus. All of this is incompatible with minimalist approach, which, as I have defined it, is that the shoe should do only what is necessary for a shoe to do, and the foot should be left free to do the rest. It would take a longer book than this to adequately treat the pros and cons of heeled and flat shoes in all common uses. For this reason, I have not included flatness as a performance specification. Suffice to say that flatness is integral to the minimalist philosophy of the isodynamic shoe design.

Part 3

The Shoes in Action

Now that we have performance specifications for shoes, we can see how the isodynamic shoe design satisfies the specifications. In this Part, I will show how the construction and materials of the shoe enable it to meet the specifications.

Chapter 10

Flatness and Flexibility

A key challenge for the isodynamic shoe design is to imbue a flat shoe bottom with all the desirable properties of a walking surface. We will see in this chapter how flatness and flexibility interact to meet the challenge.

Static Conditions

The concept of flatness may be considered under static and dynamic conditions. "Dynamic" refers to when the wearer is walking, running, etc. "Static" is the rest of the time (Watkins, p. 211).

There are two aspects of static conditions: weight-bearing and non weight-bearing. Weight-bearing occurs when the wearer is standing or has at least one foot on the ground while sitting. Under weight-bearing conditions, the shoe should lie flat under the foot.

Under non weight-bearing conditions, the toes may tend to hyperextend, *i.e*, bend upward, slightly. It is therefore acceptable for the front of the shoe to have some curvature (toe spring), provided that: (1) the shoe doesn't force the toes back further than they would naturally bend; and (2) the shoe flattens without resistance in the weight-bearing condition.

Dynamic Conditions

The dynamic aspect of flatness is more complicated than the static. We want two things: (1) that there be a flat bottom under every part of the foot that is bearing weight; and (2) that the height of the bottom be uniform over the entire length and width of the foot. These conditions would be easy to achieve if the bottom of the shoe could be made of a thin, flexible material, like the leather of an

29

Indian moccasin. The problem with thin materials, though, is that they don't provide shock absorption or thermal insulation. To achieve those qualities, we have to use materials that have a certain thickness--at least one-half-inch thick in my experience--and density. Such materials are not perfectly flexible, however, and so we are forced to consider tradeoffs.

The Sole

Under the forefoot, a minimum limit on flexibility is set by the need for the sole to bend at the base of the toes as the foot moves into and through the push-off phase. This is illustrated in Figure 4. The sole must bend to an angle of approximately 55-60°. Some muscle energy is used in bending the sole, so we want the sole to be as flexible as possible without being too much less firm than the material under the heel of the foot. I have obtained good results with a thin rubber outsole and a one-half-inch thick EVA foam midsole with a Shore A hardness of approximately 35 and a resilience of 25%. As with all materials described in this book, the specification of rubber and EVA is not necessarily the last word on the subject.

An otherwise flexible sole will be unable to flex if the toe area of the upper is rigid. Modern shoes have a stiff

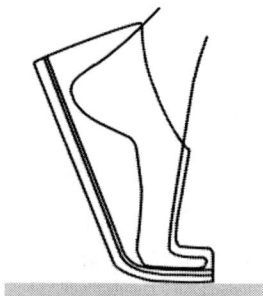

Figure 4. Isodynamic shoe in push-off phase of gait cycle. Sole bends to accommodate hyperextension of toes.

30

reinforcement between the shell and lining (the "box toe") that allows flexion only in the region around the ball of the foot. This requires the shoe to have a toe spring to keep the wearer from going up on the tip of his toes in the push-off phase. A flat shoe does not have a toe spring, so it needs flexibility in the entire toe area. The toe area of the upper can have a reinforcement to maintain its shape, but it must be flexible.

The Heel

As the heel of the shoe touches down at heel strike, and the foot begins to bear down, from back to front, the heel of the shoe must bend in order to maintain a flat surface under the weight-bearing area of the foot. This is illustrated in Figure 5. The tradeoff between flexibility and shock absorption is different here than in the sole. The heel does not have to bend as sharply as the sole. More shock absorption is needed in the heel than the sole because the impact is stronger in the heel. This dictates that a denser material be used for the heel than for the sole. I have obtained good results with a thin rubber outsole and a one-half-inch thick EVA foam midsole with a Shore A hardness of approximately 45 and resilience of 25%. These, it may be recalled, are the hardness and resilience specifications for the ideal walking surface.

As with the sole, a flexible heel requires a flexible upper. Figure 5 shows how a fold, or dimple, forms in the

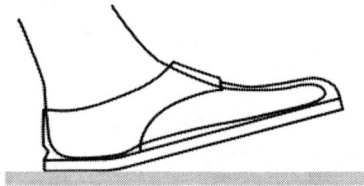

Figure 5. Isodynamic shoe at heelstrike. Heel lies flat under foot.

back of the upper. In contrast, the modern shoe has a stiff reinforcement, called a "heel counter," to maintain the shape of the backpart.

The two illustrations above only show flexion perpendicular to the direction of motion, but an isodynamic shoe can flex in any orientation, and so can accommodate the transverse and oblique axes of the foot previously discussed. The sole of a modern shoe, as illustrated in Figure 12, rolls forward in a predetermined orientation and has only limited flexion in other orientations.

Longitudinal Torsion

As noted in the chapter on biomechanics, the forefoot, midfoot and hindfoot do not rotate as a unit: a certain amount of longitudinal torsion is normal in barefoot walking. This normal torsion is different from the excessive twisting that results in injury. One of the worst features of the modern shoe is its torsional stiffness. Because motions of some joints in the feet are coupled to the rotational position of either the forefoot or the hindfoot, torsional stiffness reduces the range of motion of those joints. This may be one reason that many people appear awkward when walking in conventional shoes.

There are basically two ways a shoe can accommodate normal longitudinal torsion. The first is by having a "split sole"--a stiff sole under the forefoot and a stiff heel, with a soft bridge between them. This construction is employed in certain dancing shoes, as illustrated in Figure 6. The front and back of this shoe rotate separately. The

Figure 6. Dancing shoes, featuring split sole.

dance world understands that torsional stiffness in a shoe is incompatible with free movement. Some models of athletic shoes have a type of split sole, where instead of a complete gap the sole becomes very narrow in the middle.

The second way is for the backpart of the upper to be flexible. This allows the hindfoot to rotate freely inside the shoe, irrespective of the rotation of the forefoot or midfoot. In this way, the shoe resembles slippers and sandals, which do not restrain rotation of the hindfoot. (This may be the main reason why slippers and sandals are more comfortable than regular shoes.) This second method makes it possible to have a full sole along the bottom, without gaps or bridges, which is an advantage over the split sole in that it provides a complete walking surface under the weight-bearing areas of the foot. A footprint made by a bare foot is continuous along the outer edge, as illustrated in Figure 7. Some weight is borne by each area of the foot shown in the illustration. If the shoe has a gap in any of those areas, weight has to be shifted to some area of the foot ahead of or behind the gap.

The isodynamic shoe uses the second method: a flexible upper to accommodate the normal longitudinal torsion of the foot.

The discussion of flexibility highlights one overall difference between an isodynamic shoe and a modern shoe. The upper of an isodynamic shoe functions essentially as a strap to keep the bottom of the shoe from flying away from the foot. The upper of a modern shoe functions in part as a brace against flexion of the bottom.

Figure 7. Footprint. Weight bearing is continuous along outer edge.

Chapter 11

Shock Absorption

As we saw in the discussion of the hardness of walking surfaces, a balance has to be drawn between generation of ground reaction forces, kinesthetic sense, and shock. This results in a surface that is relatively hard, but with some resilience. In translating the concept of hardness to shoes, it is convenient to speak of "shock absorption." I use the word "shock" in the subjective sense: the feeling of a sudden impact. This is consonant with The Complete Footwear Dictionary's definition of "step shock" as "[t]he jolt effect occurring with each step or stride when walking on a nonresilient surface with shoes lacking shock absorption."

There are two primary methods for reducing the perception of heel shock. One method is to put shock-absorbing materials under the heel. The other method is to deflect the impact away from the heel. Shock absorption can be attained by the use of materials of appropriate hardness, resilience and thickness. My experience has been that a one-half-inch thick EVA foam midsole with Shore A hardness of approximately 35 under the forefoot and 45 under the heel produces good results. It provides just enough shock absorption to prevent heel pain. The trade-offs involved in selecting the proper hardness of the material were discussed in the chapter on flatness and flexibility.

Shock deflection is a feature of modern shoes. This is illustrated in Figure 8. The stiff materials and structure of the shoe diffuse the force of impact across the parts of the foot that are in contact with the shoe. As a result, the impact to the bottom of the heel is diminished. Less heel shock is felt. Something similar may occur with high-heeled shoes. Athletic shoes may feature a combination of shock absorption and shock deflection.

The disadvantage of shock deflection is that it re-duces kinesthetic awareness. The wearer doesn't feel how hard his feet are coming down. The full ground reaction force is still transmitted into the body, but the body may not optimally adjust itself to handle that force.

Isodynamic shoes feature shock absorption obtained by using shock-absorbing materials. The shock deflection method used in modern shoes requires stiff construction and a raised heel, and is therefore incompatible with the flatness and flexibility of the isodynamic shoe.

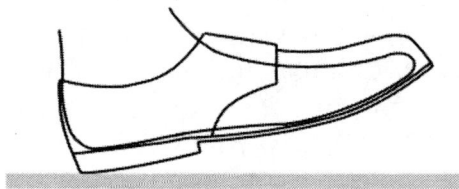

Figure 8. Modern shoe at heelstrike.

Chapter 12

Internal Traction

In the chapter on walking surfaces, we considered the importance of friction between the foot and the surface beneath it. In this chapter, I will use the word "traction" to refer to resistance to slipping. To "have traction" means not to slip. By "slipping," I mean uncontrolled or excessive sliding. This terminology is less precise than measuring coefficients of friction, but it is sufficient for our purposes. Control of slipping is crucial to shoe design. Slipping results in wasted energy and possible strains of the foot. On the other hand, it would be difficult to walk if the foot adhered too tightly to the shoe, especially if the outsole of the shoe has a very high coefficient of friction with the ground. There must be some minimal amount of sliding between the bottom of the foot and the shoe, but it has to stop short of the point of slipping.

There are two primary methods to prevent the foot from slipping inside the shoe. The first is to create traction between the insole and the bottom of the foot. The second is for the shoe to hold the foot in a way that restrains its forward and backward motion. The second method is accomplished by a combination of a tight upper and the shape of the insole.

Traction is the simpler method. It has been used in footwear since the time of the earliest shoes. It works satisfactorily with bare feet because the feet have natural traction against leather and other materials historically used in footwear. Socks complicate the issue. The fabrics commonly used in making socks have a lower coefficient of friction than skin. There can be slippage between the sock and foot and between the sock and the insole. Affirmative measures are therefore needed in order to maintain traction in shoes that are to be worn with socks. I have found that good (if not ideal) traction can be achieved by coating the insole with a material like RTV silicone rubber and drawing lateral grooves in the material before it cures. An incidental benefit of a grooved or textured insole surface is an increase in sensory awareness. In the future, one hopes, socks will be designed to give proper traction within a shoe.

Modern shoes use a combination of tight uppers and shaping of the insole to keep the foot from slipping. If the shoe fits snugly across the instep when it is laced up, the foot will not slide forward. If the backpart (with its heel counter) fits snugly, the foot will not slide backward.

Tight uppers are used to pernicious effect in women's shoes. If the slope of the shoe is steep enough, the toes must be held very tightly by the vamp to keep the foot from sliding down. The vamp must therefore be narrower than the foot. This is a structural necessity, not just a matter of fashion. The consequence for women who wear these shoes is bunions and other foot pathology (Menz).

The insole of a modern shoe (see Figure 2) restrains motion through its shape. In the shank area, it is convex and it fits into the concavity of the foot's longitudinal arch. The shoe thereby locks the foot in place. The effect is increased when the shoe has an "arch support."

Isodynamic shoes rely on traction and good fit across the instep to prevent slipping within the shoe. They do not need to use tight vamps, stiff heel counters and arch supports. In this respect, internal traction is the linchpin of the isodynamic design because it makes everything else possible.

Chapter 13

Stability

S tability is one of the desired qualities of a walking surface. The isodynamic design achieves stability through flexibility: the sole is sufficiently flexible so that the bottom always lies flat under the part of the foot that is bearing weight, and the upper is sufficiently flexible so as not to stop the bottom from flexing. Figures 4 and 5 illustrate how this works during the push-off and heel strike phases of the gait cycle.

The modern shoe does not provide a stable walking surface. As shown in Figures 8 and 12, the shoe pivots around a point near the back of the heel, and rolls over the front part of the sole.

Chapter 14

External Traction

External traction is the traction of the outsole against the ground. There is a large body of research on this subject. Industry standards for acceptable coefficients of friction for flooring and outsole materials have been published. The impetus for this activity is, presumably, concern over liability for slip and fall accidents.

Standard outsole materials are satisfactory for everyday walking on sidewalks and floors in isodynamic shoes. Special materials and tread patterns may be needed to prevent sliding on very slippery surfaces, like mud and oil. Sometimes a little sliding is desirable: bowling shoes, for example have soles with a low coefficient of friction.

I have not included external traction among the performance specifications, because, although it is indispensable, it is not usually a problem.

Chapter 15

Environment

Shoes surround the feet. They create a discrete environment and, to a degree, insulate the foot from the outside environment. In the footwear industry, the art of optimizing the inner environment through the use of special materials or other methods is known as "climate control."

The inner environment of isodynamic shoes is more a function of materials than design. There may be at least one inherent environmental difference between isodynamic and modern shoes. In isodynamic shoes, the arch is free to rise and flatten during the gait cycle, unhindered by an "arch support." This may result in a bit of pumping action that increases the air flow within the shoe.

Chapter 16

Protection From the Elements

Isodynamic shoes protect the feet from the outside environment in much the same way as modern shoes. The half-inch-thick EVA midsole, previously mentioned, provides thermal insulation and a barrier against penetration by sharp objects.

Chapter 17

Support

If there is one virtue that is most often extolled in shoes, it is support. You must wear shoes that support your feet—so we are told by parents, health professionals and footwear manufacturers. What, then, is wrong with our feet, that they need to be supported by shoes? In this chapter, I will examine the concept of support and ask whether it is really necessary.

To begin with, we may note that there are two applicable senses of the word "support." In the first sense, we can say that shoes support the feet by providing a favorable environment. Support, in this sense, is nothing more than the sum of all the shoe's functions. I did not list this kind of support as a performance specification because it has no specific meaning. I would not object to using the word in this way if it were not so easily confused with the other sense of the word.

The second sense of support involves the arches of the foot. The normal foot has two longitudinal (front-to-back) and one lateral (side-to-side) arches (Watkins, p. 85). An arch of the foot is not a particular piece, like the heel bone, but rather a geometrical arrangement of a portion of the foot. If you stand on a flat surface, the arches are the areas where the bottom of the foot does not touch the surface. To state it another way, the arches are areas where the shape of the bottom of the foot is concave rather than flat. The arches are formed by the arrangement of the bones of the foot, somewhat like architectural arches are formed by arrangements of stones. Unlike the arch of a building, however, the arches of the feet are not rigid. It is normal for them to flatten temporarily during weight-bearing phases of the gait (or running) cycle (p. 211). Geometrically, this is equivalent to saying that it is normal for the foot to stretch during weight-bearing phases of the gait cycle, because a curved object straightens when you pull its ends apart. You can see this for yourself in the following way: Stand barefoot on a flat surface with most of your weight on the left foot. Carefully shift as much of your weight as you can, without losing your balance, to your right foot while keeping your right heel in place. You will notice that the arch of the right foot flattens somewhat and the large toe moves forward slightly. You will also notice that your right foot rolls inward to some extent. This rolling is known as pronation.

Because arches are concave, it is possible to put a convex object under the arch which will prevent the arch from flattening. This is what an arch support is. An arch support is an insert or a portion of the insole which fits under the main arch of the foot, and which, by its shape and composition, prevents or restricts the arch from flattening. The degree to which it restricts flattening depends on the precise shape of the arch support in relation to the shape of the arch. If it fits exactly into the arch, and if it is sufficiently rigid, it will completely prevent the arch from flattening. In the process, it will restrict pronation of the foot. I believe this is what most people, including orthopedists and podiatrists, have in mind when they speak of shoes supporting the feet.

It should be obvious from the above that there is something problematical about arch supports, because they interfere with the normal functioning of the feet. However, I don't want to jump to the conclusion that arch supports are always bad. I am not criticizing the medical use of arch supports, or other devices, under conditions where it may serve a therapeutic purpose to restrain flattening of the arches, or stretching or pronation of the feet. Moreover, I can think of some nonmedical circumstances where arch supports can be beneficial. As I noted in Chapter 12, arch supports can restrain the foot from sliding inside heeled shoes. Another function of arch supports is that they distribute the weight borne by the foot over a larger area, thereby reducing pressure on the heel and forefoot. This can be especially important in heeled shoes, because raising the heel of the foot shifts weight to the forefoot. Distributing weight can be beneficial for people who spend a lot of time standing. For these reasons, and because there are many complexities in the subject, I do not want to make a categorical judgment about arch supports or shoes that support the feet in the strong sense of the word.

The human foot is the result of adaptation, over long eons, by barefoot creatures who had nothing to support their feet except the foot's own internal structure. Humans walked and ran barefoot or in simple moccasins until comparatively recently. Running is less stressful than walking, and barefoot runners fare as well or better in terms of foot health than runners who wear modern running shoes (Driscoll). Thus if a shoe can create conditions similar to barefoot walking, there is no reason to suppose it will cause pathology in healthy people, notwithstanding an absence of arch support.

Chapter 18

Comfort and Fit

Comfort is something people look for in a shoe. But what does it really mean for a shoe to be comfortable? And is comfort an unqualified good?

To most people's minds, a comfortable shoe is one that does not cause pain. There are several ways in which shoes can cause pain. The most obvious is poor fit. This usually means that the shoe is too tight. Sometimes it's the consumer's fault: people buy shoes that are too small because they don't want their feet to look big. Many people, however, simply have difficulty finding ready-made shoes that fit properly. Shoe size is measured in two dimensions: length of the foot and width across the ball of the foot. Two people may have feet of the same size in those dimensions, but their feet may differ greatly in a number of other dimensions. The last that a manufacturer uses represents an unscientific approximation of a typical foot of a given size: shoes made on that last may not fit anyone exactly. To make matters worse, many ready-made shoes, particularly athletic shoes, don't come in widths. Thus it is that mass production yields poorly fitting shoes.

Even custom shoes, hand made on custom lasts by the finest craftsmen, can be uncomfortable. This, I believe, is a consequence of the essential nature of the modern shoe, particularly its lack of neutrality and restriction of range of motion. (It may also result from the fact that custom shoemakers build their lasts from preshaped forms rather than a cast of the customer's feet.) Some people have stances and gaits that are compatible with modern shoes. For others, their shoes are always fighting with their bodies.

Women's high-heeled shoes seem designed to cause pain. This is a subject that belongs to the social sciences as much as to medical science and shoe engineering.

With isodynamic shoes, there is no pain caused by lack of neutrality or restriction of motion--for the simple reason that the shoes are neutral and do not restrict range of motion. Whether they fit as well as modern shoes is an open question. The shoes I have produced have been made with custom lasts based on a cast of the foot, so it is inevitable that they fit well. If the design were mass-produced, proper fit could be an issue. But ready-made isodynamic shoes may have more leeway than modern shoes: because of internal traction, isodynamic shoes have to fit snugly only over the instep. They can err on the side of being too large in all other dimensions. This may allow a single shoe size to fit a variety of differently shaped feet.

There is a point where lack of pain shades off into lack, or reduction, of normal sensation. Customers may consider a shoe comfortable because it reduces the normal sensations that a person feels in walking or standing. This type of "comfort" can result from: (1) shock absorption beyond the optimal level; and (2) an insole that closely fits the contour of the sole of the foot, so that weight is evenly distributed over the insole. There is evidence that people equate comfort with evenness of pressure across the entire bottom surface of the foot (Jordan). But reduction of sensation can mean reduction of the kinesthetic awareness inherent in barefoot walking. This may cause alterations of the gait, which eventually may result in discomfort in various parts of the body. A net long-term discomfort may thereby result from wearing apparently comfortable shoes.

Aside from the issue of pain, there is an esthetic, even philosophical, issue of how much sensation a person wants to feel through his feet when walking. Do you want to feel the ground beneath you? Do you prefer the floating feeling of high-tech athletic shoes? The aesthetic of the isodynamic design is that walking in shoes should feel, as much as reasonably possible, like walking barefoot. It does not seek a softer comfort.

Chapter 19

Appearance

A shoe is an item of apparel as well as a medium for transmission of forces. How a person feels in a pair of shoes is, in part, a function of how the shoes look. While the focus of this book is on design as engineering, there is also the esthetic side of shoe design. I will limit my comments, however to just a few esthetic issues directly related to the isodynamic shoe.

Isodynamic shoes are flat. People who want the extra height they get from heels will have to accept this fact. On the other hand, isodynamic shoes do have some height because of their thick midsole.

The shoes have a high front. They also tend to have a fairly broad-shaped toe, depending on the individual. The reason for this is explained in the chapter on toe shape.

The shoes tend to develop a dimple in the back, as a result of flexing at heel strike. See Figure 5. This isn't very noticeable, especially if the person is wearing pants. It can probably be minimized by using appropriate reinforcing materials in the backpart.

These qualities may be seen by some as negatives from a visual viewpoint. On the positive side, the wearer should have a more graceful and buoyant walk and better posture, than someone wearing modern or high-heeled shoes. The fashion challenge in making isodynamic shoes is to bring out the positive dynamic qualities while either disguising the negatives or transforming them into a fashion of their own. In women's shoes, in particular, the esthetic must be one of lightness and motion--in contrast to the vertical esthetic of styles like stiletto heels.

Chapter 20

How Isodynamic Shoes Work

The defining features of isodynamic shoes are that they are flat-soled, flexible in the uppers and bottom, and have good shock absorption and internal traction. As we have seen in the preceding chapters, they satisfy the performance specifications for general-purpose walking shoes in the following ways:

1. *Provide a walking surface of optimal hardness, contour, coefficient of friction and stability.* Isodynamic shoes do this by having a sole that is level, and that strikes a golden mean between flexibility and shock absorption. The insole is made of, or covered with, materials having the right coefficient of friction and texture to give traction when the wearer is wearing socks.

2. *Allow full range of motion of all parts of the body, particularly the foot and ankle.* Isodynamic shoes do this through flatness and flexibility in the sole and uppers.

3. *Be neutral.* Isodynamic shoes do this, too, through flatness and flexibility in the sole and uppers.

4. *Optimize kinesthetic awareness.* Isodynamic shoes do this by having a sole with enough softness and resilience to absorb shock, but not so much that it takes away the feel of the ground.

5. *Protect the feet from the outside environment.* Isodynamic shoes do this in much the same manner as other shoes. The midsole of isodynamic shoes provides thermal insulation and shields the foot from sharp objects.

6. *Provide a healthy internal environment for the feet.* Isodynamic shoes are comparable to other shoes in this respect.

7. *Fit well.* This is more a function of the last than the shoe architecture. Mass-produced isodynamic shoes may have an advantage in that they need only be snug over the instep, and can be slightly loose in the rest of the shoe.

8. *Look good.* This is a function of styling and materials.

49

All in all, the elements of the isodynamic design work together to provide an experience that approaches the natural process of barefoot walking.

Part 4

Some Parts of the Shoe

The previous two Parts covered the basic performance criteria of shoes. This Part will take up a set of issues that are peripheral to the concept of the isodynamic shoe, but are still important in determining comfort and wearability. Each of the next chapters is devoted to a different part of the shoe, ending with a chapter on a quasi-part: socks.

Chapter 21

Toe Shape

Toe shape or "toe expression," is defined as "the shape of the last or shoe at the toe." The toe is the most visible part of the shoe. Different styles of shoes have differently shaped toes, from the pointed toe of the Western boot to the round toe of athletic shoes and work boots. In designing a toe shape, we can attempt to follow the shape of the foot or aim for something else. In this chapter, we will first look at toe shape in the horizontal plane (as viewed from the top) and then in the vertical plane (as viewed from the side).

Horizontal Plane

For the foot to fit inside the shoe without being squeezed, the toe of the shoe has to be at least as wide as the toes of the feet at their widest point, which is at or just behind the base of the toes. If we want a toe shape that approximately follows the shape of the foot, we draw the front of the shoe about one-quarter inch from the foot at its closest point. This is illustrated in Figure 9.

The one-quarter-inch-plus margin is called a "fitting allowance." For a more rounded and bilaterally symmetrical toe shape, the arc of the front edge can be lengthened

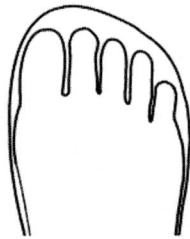

Figure 9. Toe shape, approximating shape of the foot.

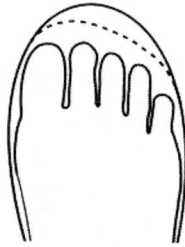

Figure 10. Toe shape, rounded.

and centered as illustrated in Figure 10. The dotted line represents the front edge of Figure 9. For a more pointed toe, the front edge can be further extended and narrowed as illustrated in Figure 11.

Notice that the more pointed the toe, and the more centered the apex of the front edge, the further forward the edge has to extend. This creates, in effect, an extra extension of the sole and the upper, and raises the questions: (1) how should this extension be constructed; and (2) should there be such an extension at all in an isodynamic shoe?

The advantage, or perhaps, the necessity of the extension in modern shoes is that it is required by the toe spring. As illustrated in Figure 12, the foot rolls over the toe spring during the toe off period of the gait cycle. The toe extension has to be long enough so that the shoe does not roll over its front edge. This, in turn, requires the ex-

Figure 11. Toe shape, pointed.

tension to be rigid. It has to maintain its shape during toe off.

One disadvantage of the extension is that it adds to the risk of the shoe's colliding with or being run over by something. It's a hazard, but perhaps not a significant one.

In an isodynamic shoe, there is no toe spring. There is thus no need for an extension. Moreover, a flexible extension would tend to give way in situations where weight is put on the toe as the foot comes down, such as when climbing up stairs. That would be a hazard. It follows that an isodynamic shoe should not have a forward extension, and the toe shape should be close to that of the foot.

Vertical Plane

The prime issue in designing the vertical aspect of the toe shape is how much height the toes require. We have seen that the toes hyperextend during the push-off and swing phases of the gait cycle. The toe shape has to be high enough to accommodate this; otherwise the toes can become deformed. Because the isodynamic shoe does not have a toe extension, there is no opportunity to taper down the height toward the front, and so the toe necessarily has a somewhat boxy appearance. The toe can taper laterally, since the small toes need less height than the large toe.

Modern shoes can have a slimmer appearance than isodynamic shoes. The height can taper downwards toward the front in the extension, since the extension is empty. Also, depending on the geometry of the toe spring, the shoes may keep the toes permanently hyperextended, to some degree, above the toe spring. The top of the shoe can be high enough for hyperextension, but the shoe will be thin because the bottom curves up above the ground.

High-heeled shoes keep the toes in hyperextension to a significant degree. They thereby maintain a low profile in the toe, but at a price of discomfort and injury.

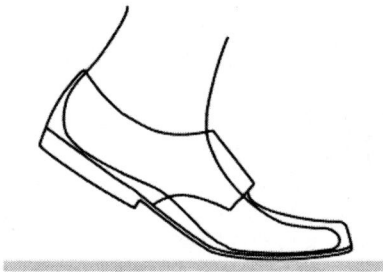

Figure 12. Modern shoe at push-off.

Chapter 22

Toe Box

The toe box is defined as the "firm, reinforced toe area of a shoe." The "box toe" or "toe puff" is the reinforcing material placed between the lining and the shell of the toe box.

In an isodynamic shoe, the box toe is made of flexible material that easily bends and returns to its original shape. This enables flexing of the sole and upper in the toe area. Modern shoes, in contrast, need a fairly rigid box toe to maintain the shape of the toe spring.

Chapter 23

Heel Shape

How far back the heel should extend is determined by the geometry of the shoe at heel strike. As illustrated in Figures 3 and 5, the heel of the isodynamic shoe extends somewhat behind the back edge of the foot. In modern dress shoes, as shown in Figure 8, the back edge of the heel is ahead of the back edge of the heel of the foot. Athletic shoes normally have a heel that extends in back of the foot. Unlike the isodynamic shoe, they normally have a stiff construction in the backpart above the heel.

Chapter 24

Eyelet Tabs

The eyelet tabs are the part of the edges of the quarters that close over the tongue. They hold the eyelets (or "lace holes" if there are no eyelets). This chapter considers the question of their proper length in those styles that have eyelet tabs.

In isodynamic shoes, a "vamp crease" forms in the upper, somewhat behind the base of the toes. You don't want this crease running through the eyelet tabs. Therefore, the front edge of the tabs should be behind the crease. Another crease will form at the junction of the ankle and midfoot if the eyelet tabs and tongue go up that high. That would be uncomfortable because the crease would dig into the foot. Thus the eyelet tabs and tongue should not extend that far back. The two crease lines set the limits for the length of the eyelet tabs. In a men's shoe of medium size, this results in eyelet tabs about an inch and three-quarters long. The tabs will hold four eyelets.

If the shoe fits properly, the eyelet tabs will lie directly over the instep, or midfoot, and when laced up will apply a slight, even pressure to the foot. In that way, they will hold the shoe securely on the foot.

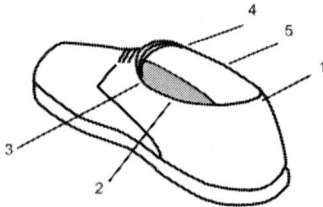

Figure 13. Top line.

58

Chapter 25

Top Line

The top line of a shoe comprises the upper edge of the two quarters. A close look at the top line reveals many subtleties about the shoe. The top line is a three-dimensional curve. We will view it first in the vertical plane and then in the horizontal.

Vertical Plane

The top line in a blucher is defined by five points, as illustrated in Figure 13. Point 1 is at the top of the back seam. (The drawing does not actually show the seam, but it would be slightly to the left of the back edge as seen in this perspective.) Points 3 and 4 are at the corners of the eyelet tabs (also not shown). Points 2 and 5 are determined by the heights of the medial and lateral malleoli--the inner and outer knobs of the ankle. Normally the medial malleolus is higher than the lateral. You don't want the shoe cutting into the malleoli, so the top line should pass under them. There is no fixed measure for how far under the malleoli the top line must pass.

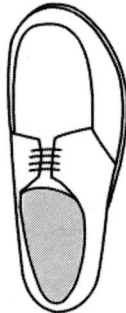

Figure 14. Top line of modern shoe.

As noted in the chapter on eyelet tabs, Points 3 and 4 should be slightly in front of the junction of the ankle and the midfoot. The standard height of Point 1 is about two and a half to three inches measured along the back line. Upper and lower boundaries are set by the kinematics of the ankle: if Point 1 is too high, the shoe will tend to dig into the ankle on plantarflexion (bending of the foot downward). If it is too low, it will tend to get stretched on dorsiflexion (bending of the foot upward).

Within the limits imposed by the five points, there is some room to vary the shape of the line. To keep the line tight, the overall curve has to be kept simple. The main ways of varying the curve are by playing with the heights of Points 2 and 5, and possibly moving them slightly forward or backward.

Horizontal Plane

The vertical shape of the top line results from a design choice about the top line itself. The horizontal shape, on the other hand, follows from the shape of the entire back of the shoe.

Figure 14 is a modern shoe. The illustration is based on a custom shoe that was made on what I would call a semi-custom last. The last starts out as a preshaped wooden form: it is adjusted by adding patches of leather in some places and sanding down the down in other places.

Figure 15 is an isodynamic shoe made on a fully custom last. The first stage in making the last was to make a cast of the foot. The top line, therefore, exactly matches the perimeter of the foot at that height. The difference between the two top lines is readily apparent. The top line of Figure 14 is narrow, approximately elliptical and has the same longitudinal axis as the entire shoe. The top line of . Figure 15 is an irregular shape. It is wider than Figure 14, so much so that the part of the medial half of the top line lies outside the line of the sole. Its long axis is at an angle to the axis of the shoe.

The top line in Figure 15 expresses the complex geometry of the foot, as may be seen, for example, in an MRI view of the ankle (Stoller, p. 566). The top line in Figure 14 is more about last making and shoemaking than about the foot: a symmetrical last is easier to make and easier to work leather on.

Figure 15. Top line of isodynamic shoe.

Chapter 26

Socks

Socks are an X-factor in shoe performance. They affect two of the cardinal properties of the walking surface: hardness and friction. Socks can provide a certain amount of shock absorption, depending on thickness and fabric. They can significantly reduce internal traction.

Socks also affect the climate within the shoe. They absorb moisture. They may reduce the airflow within the shoe. They usually increase the thermal insulation of the shoe. They thereby contribute to keeping the foot dry and warm, sometimes too warm.

Finally, socks, especially thick socks, make the shoe fit more tightly. Thick socks make it possible to wear what would otherwise be a too-large pair of sneakers or running shoes.

We have reached the end of our journey through science and shoe design. I hope you have found something useful along the way. In concluding this book, I want to emphasize that bad shoes are not inevitable. If enough people understand what they need from shoes, we may one day enter a world of shoes that don't hurt.

References

Aristotle. Progression of animals. In *The Loeb Classical Library: Aristotle Vol. 12.* 1961. Cambridge, Mass.: Harvard University Press.

Baez, J. 1963. On *Joan Baez in Concert Part 2.* [2002 CD]. Santa Monica, Cal.: Vanguard Records.

Collins, J.J. 1996. Interviewed in S. Blakeslee, Vibrating soles help people regain balance. *New York Times,* Jan. 12, 2006, p. F2.

Driscoll, D.G. 2004. Barefoot running: a natural step for the endurance athlete. *Track Coach* 168. Reproduced at http://nhscc.home.comcast.net/l3_paper.htm.

Encyclopedia of Physics (2d ed.). 1991. N.Y.: VCH Publishers, Inc.

Jordan, C., C. Payton and R. Bartlett. 1997. Perceived comfort and pressure distribution in casual footwear. *Clin. Biomech.* 12: S5.

Menz, H.B. and M.E. Morris. 2005. Footwear characteristics and foot problems in older people. *Gerontology.* 51:346-351.

Muybridge, E. 1995. *The Human Figure in Motion.* Mineola, N.Y.: Dover Publications, Inc.

Peacock, J. 2005. *Shoes: The Complete Sourcebook.* London: Thames & Hudson, Ltd.

Priplata, A.A., et al. 2006. Noise-enhanced balance control in patients with diabetes and patients with stroke. *Ann. Neurol.* 59:4-12.

Robbins, S., E. Waked, P. Allard, J. McClaran and N. Krouglicof. 1997. Foot position awareness in younger and older man: the influence of footwear sole properties. *J. Am. Geriatr. Soc.* 45: 61-66.

Robbins, S., E. Waked, and N. Krouglicof. 1998. Improving balance. *J. Am. Geriatr. Soc.* 46: 1663-1670.

Stoller, D.W. 1999. *MRI, Arthroscopy, and Surgical Anatomy of The Joints.* Philadelphia: Lippincott Williams & Wilkins.

Surburg, P. 1995. Flexibility training: program design. In P. Miller, ed., *Fitness Programming and Physical Disability.* Champaign, Ill.: Human Kinetics.

The Complete Footwear Dictionary (2d ed.). 2000. Malabar, Fla.: Krieger Publishing Co..

Thompson, D.M. 2005. *Ground Reaction Force.* http://moon.ouhsc.edu/dthompso/gait/ kinetics/GRF-BKGND.HTM

Vass, L. and M. Molnár. 1999. *Handmade Shoes for Men.* Cologne, Germany: Könemann Verlagsgesellschaft mbH.

Watkins, J. 1999. *Structure and Function of the Musculoskeletal System.* Champaign, Ill.: Human Kinetics.

Whittle, M.W. 1996. *Gait Analysis (2d ed.).* Oxford: Butterworth-Heinemann.

Zatsiorsky, V.M. 1998. *Kinematics of Human Motion.* Champaign, Ill.: Human Kinetics.

ISBN 141209706-1